ALTERNATOR BOOKS™

WHO ELSE IN HISTORY?

Hidden Heroes in TECHNOLOGY

Carol Kim

Lerner Publications ◆ Minneapolis

TO MY DAD AND BROTHERS, MY FAVORITE TECH EXPERTS

Lerner Publications Company
An imprint of Lerner Publishing Group, Inc.
241 First Avenue North
Minneapolis, MN 55401 USA

For reading levels and more information, look up this title at www.lernerbooks.com.

Main body text set in Aptifer Sans LT Pro.
Typeface provided by Linotype AG.

Library of Congress Cataloging-in-Publication Data

Names: Kim, Carol, author.
Title: Hidden heroes in technology / Carol Kim.
Description: Minneapolis : Lerner Publications , [2023] | Series: Alternator books. Who else in history? | Includes bibliographical references and index. | Audience: Ages 8–12 | Audience: Grades 4–6 | Summary: "Tech is everywhere! Learn more about the innovators who created the technology world. These innovators are women, people of color, and members of the LGBTQIA+ community and ready to be recognized"— Provided by publisher.
Identifiers: LCCN 2021045392 (print) | LCCN 2021045393 (ebook) | ISBN 9781728458441 (lib. bdg.) | ISBN 9781728464039 (pbk) | ISBN 9781728462578 (eb pdf)
Subjects: LCSH: Inventors—Biography—Juvenile literature. | Minority engineers—Biography—Juvenile literature.
Classification: LCC T39 .K46 2023 (print) | LCC T39 (ebook) | DDC 609.2/273—dc23/eng/20211118

LC record available at https://lccn.loc.gov/2021045392
LC ebook record available at https://lccn.loc.gov/2021045393

Manufactured in the United States of America
1-50873-50210-2/15/2022

TABLE OF CONTENTS

ADVANCING TECHNOLOGY

Many stories of people behind important advances in technology are just beginning to be told. People like Ada Lovelace, who was the world's first computer programmer in the 1800s, and Grace Hopper, who developed some of the first modern computer programming languages, are finally receiving recognition.

Lovelace and Hopper helped make modern computers possible.

From the invention of color television to how we use the internet, people from underrepresented groups have played a key role. These innovators completed groundbreaking work that made possible many modern devices and tools that we use every day.

Lovelace discovered that a computer could follow a program, or a series of instructions.

Hopper (*second from right*) works on an early commercial computer in 1960.

PROBLEM SOLVERS

Inventions usually start with a problem. The following inventors came across a problem and decided to fix it. Their inventions have greatly improved people's lives.

Granville T. Woods improved communication by inventing the induction telegraph.

Woods created a company to develop and build his inventions.

Granville T. Woods

Granville T. Woods was a creative inventor. He registered about sixty patents over his life.

Born in 1856, Woods was fascinated by electricity. During his early twenties, he moved to New York City. He worked during the day and took classes at night. Since Black people were not allowed to use the city's library then, he had to ask friends to get books for him.

Woods improved long-distance communication with his invention of telegraphony. It allowed both voice and telegraph messages to be sent over the same line. The invention was so successful, he sold the patent to Alexander Graham Bell.

Woods also invented the induction telegraph. It is a communication system that allowed train engineers to talk with one another, which prevented many accidents.

Alan Turing

Alan Turing helped crack the Enigma code. In World War II (1939–1945), the German military used this code to send secret messages so they could not be read without a key. His work helped save millions of lives.

Born in 1912, Alan struggled with school. But he was interested in science. While at Princeton University, he came up with a theory for a computing device. Turing called it the Turing Machine. It became the foundation for developing modern computers.

Much of Turing's code breaking was classified, so he did not receive full recognition for his work.

Turing also explored artificial intelligence. He developed the Turing Test to see if machines could think on their own. The test asked a series of questions to a person and to a computer program. If the person running the test could not tell which answers had come from the person and which from the machine, it would prove the machine was using artificial intelligence.

A working replica of the Turing Machine

In 1952, it was illegal in Britain to be part of the LGBTQIA+ community. As an openly gay man, Turing was criminalized for his identity. He was sentenced to a year of probation, and his criminal record prevented him from continuing his code-breaking work. He died in 1954 and is an important person in the LGBTQIA+ community. Turing has been recognized as one of the most original and brilliant thinkers of the twentieth century.

Camarena studied electrical engineering before developing the color TV.

Guillermo González Camarena

Born in 1917 in Guadalajara, Mexico, young Guillermo González Camarena showed a talent for inventing. He built electric toys, and at the age of eight, he made his first radio transmitter. Four years later, he built a working radio. His most famous invention came when he was twenty-three years old—a type of color television. It was called the Trichromatic Field Sequential System.

A television in 1955

In 1942, Camarena received the first patent for a color TV in both the US and Mexico. Wanting to make his technology widely available, Camarena later developed the Simplified Bicolor System. This technology cut costs, which allowed more people to afford color televisions.

Ajay Bhatt

Ajay Bhatt was frustrated because connecting a device such as a mouse or a printer to a computer required its own special cable. If the computer and the device were made by different companies, they didn't always work together.

Bhatt was born in India in 1957, and he was hired by the technology company Intel in 1990. There he worked on creating a single connection that would work for all devices and computers, regardless of what they were or which company had made them.

Bhatt holds at least thirty patents.

A USB flash drive that works with various types of computers

In 1996, Bhatt and his team created the Universal Serial Bus (USB). It took several years before Bhatt was able to convince computer companies to use the USB. But he succeeded, and more than ten billion of these simple devices are in use around the world.

"If you have imagination, then it is possible to make your vision a reality. Don't follow the conventional path, think differently, and make the impossible possible."

—AJAY BHATT

CRITICAL THINKING

Why do you think some innovators are well known, while other creators have been overlooked?

GROUNDBREAKING PROGRAMMERS

Computers are amazing machines. People can program them to do many different tasks. One group of women was the first to program an electronic digital computer.

The first programmable and digital computer in 1945

Women of ENIAC

People were amazed when they saw a demonstration for the first electronic digital computer, the Electronic Numerical Integrator and Computer (ENIAC), in 1946. Calculations that normally took hours to do by hand could be completed in seconds.

Six women mathematicians, Kathleen (Kay) McNulty Mauchly Antonelli, Jean Jennings Bartik, Frances "Betty" Snyder Holberton, Marlyn Wescoff Meltzer, Frances Bilas Spence, and Ruth Lichterman Teitelbaum, had programmed the machine. It was a remarkable achievement. They did it with no programming tools—because none existed at the time—and little instruction.

Meltzer (*left*) and Teitelbaum (*right*) wire the ENIAC.

Yet the six women were not given credit for their work on the ENIAC. Knowledge of their contribution was forgotten for many years. Recognition came in 1997, when they were all inducted into the Women in Technology International Hall of Fame.

MATHEMATICAL TRAILBLAZERS

Many mathematicians have played important roles in technology advancements, including the following two innovators.

Annie Easley worked as a "human computer" and completed complex calculations.

Easley's code helped her research alternative energy sources.

Annie Easley

Modern hybrid cars are possible because of work done by Annie Easley decades ago. Annie was born in Alabama in 1933. Her mother told her she could achieve anything she wanted—if she put in the effort.

Easley listened to her mom. In 1955, she became one of the first Black employees at the National Aeronautics and Space Administration (NASA), which was then called the National Advisory Committee for Aeronautics. Eventually she became a NASA mathematician and computer programmer. One of Easley's most important contributions was building computer code to analyze energy sources, including the storage ability of batteries. This work led to the battery technology used in modern hybrid cars.

> "Nothing was given to minorities or women. It took some fighting to get that equal opportunity, and we're still fighting."
>
> —Annie Easley

Gladys West

A mobile device's map app uses the Global Positioning System (GPS) to make getting around easy. Gladys West's mathematical work made GPS technology possible.

In 1956, West was one of only four Black employees at the Naval Proving Ground (later the Naval Surface Warfare Center) in Virginia. She worked as a human computer and did calculations by hand. Later, she moved up to programming computers. Using computer calculations, she created a detailed and accurate model of Earth. This work was an important building block for GPS technology.

West being inducted into the Air Force Hall of Fame in 2018 for her work on the GPS

West had to wait until 2018 before her contributions were widely recognized. That year, she was inducted into the Air Force Space and Missile Pioneers Hall of Fame.

> "I felt proud of myself as a woman. . . . But as a black woman, that's another level where you have to prove to a society that hasn't accepted you for what you are. What I did was keep trying to prove that I was as good as you are."
>
> —GLADYS WEST

West (*right*) looking at data from GPS technology in 1985

CRITICAL THINKING

Why is it important for everyone to see people who look like them at school or work?

THE POWER BEHIND COMPUTERS

Computers make it possible to do many tasks more quickly and easily. But one woman's contribution allowed computers to become much more powerful than ever before.

Microchips are an important part of computers. Lynn Conway helped create microchips.

Lynn Conway

From 1956 to 2015, computing power increased by one trillion times, largely because of advances in microchip design. This stunning growth was possible because of computer scientist Lynn Conway.

Born in 1938, Conway began her career at the computer hardware company IBM in 1964. Conway had been assigned male at birth, but when she told her employers that she was going to live in alignment with her gender identity, IBM fired her.

Conway's work paved the way for modern microchips, which are found in cell phones, computers, and much more.

Conway had to start over and rebuild her career. She continued her groundbreaking work and developed a way to make it easier to design and build computer chips. Her work eventually led to the modern microchip.

Because she changed her name after leaving IBM, few knew of Conway's many achievements. In 2020 she began receiving credit for her work, including a lifetime achievement award from IBM.

INTERNET INNOVATORS

The internet makes information easy to find and share. Here are two people who helped make the internet into such a useful tool.

Radia Perlman's algorithm makes the internet reliable.

RADIA PERLMAN

Often called the Mother of the Internet, Radia Perlman believes the internet was not invented by any one person but by the work of many individuals. But Perlman's invention of the algorithm behind the Spanning Tree Protocol (STP) is why data can be reliably shared between computers.

Perlman was born in 1952. She attended Massachusetts Institute of Technology as one of only fifty women among a class of one thousand and one of the few Jewish students at the school.

Perlman has made other contributions that improved how data is transferred over computer networks. She has also written two books and holds over one hundred patents. In 2014, Perlman was inducted into the Internet Hall of Fame.

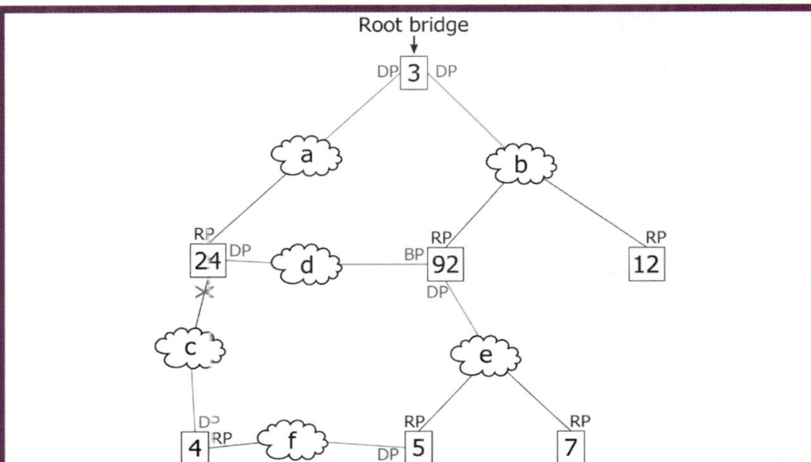

An illustration of how STP shares data

ALAN EMTAGE

Long before Google existed, Alan Emtage invented the first search engine.

Born in 1964 in Barbados, he worked as a graduate student in McGill University's computer science department. Emtage's job involved tediously searching computer files for students and faculty. To do it faster, in 1990 Emtage wrote some code to automate the task. He called it Archie.

Emtage earned two degrees in computer science from McGill University in Canada.

Users typed what they were looking for in Archie, and the program would find related files.

Emtage had created the world's first search engine. Search engines still use Archie's basic template. In 2017, Emtage was the first person from the Caribbean to be inducted into the Internet Hall of Fame.

> "I'm quite happy not to be a billionaire. . . . But [creating Archie] was done in the spirit of cooperation and realizing that while we were working on something that was big and was going to change the world."
>
> —ALAN EMTAGE

NETWORKING TECHNOLOGY

Data transfer is no longer limited to wires or cords. One man's work has allowed wireless technology to help us work in new ways.

Thomas David Petite's wireless technology sends data to devices remotely.

Thomas David Petite

Thomas David Petite was born in 1956 in Atlanta, Georgia. He grew up watching his engineer father tinker with electronics, and he was inspired to become an engineer and inventor himself.

Petite is a member of the Fond du Lac Band of the Lake Superior Chippewa Nation. He holds over one hundred patents, but he's best known for his inventions in wireless networking technology.

Petite's work directly led to the Internet of Things (IoT). Physical objects, or "things," can connect with sensors so they can exchange data over the internet. In the home, IoT allows people to control items remotely and more efficiently.

Petite is also the founder of the Native American Intellectual Property Enterprise Council, which helps Indigenous inventors and communities throughout the US.

Out of the Shadows

Brilliant, hardworking people from diverse backgrounds of race, gender, and culture have created important technology. Their accomplishments helped change the world. Let's bring these heroes out from the shadows by celebrating their contributions to the modern world.

TIMELINE

1885: Granville Woods invents telegraphony.

1936: Alan Turing invents the Turing Machine.

1942: Guillermo González Camarena receives a patent for the first color TV.

1946: Women of ENIAC program the first electronic digital computer.

Late 1970s: Annie Easley studies battery storage ability.

1980s: Gladys West performs calculations creating a detailed model of Earth.

1985: Radia Perlman invents Spanning Tree Protocol.

1990: Alan Emtage invents the Archie search engine.

1996: Ajay Bhatt invents the USB.

Late 1990s: Thomas David Petite patents wireless technology.

Glossary

algorithm: a set of guidelines for solving a problem, especially used in relation to mathematics and computer science

artificial intelligence: capability of a machine to have humanlike intelligence

human computer: a person who makes complex mathematical calculations by hand

hybrid car: a car with an engine that can run on two different fuels, usually gasoline and electricity, and can alternate between them

microchip: a tiny plate of silicon that holds many electronic parts. Microchips are used to run computers and other devices.

patent: a government grant that gives someone the right to make, use, or sell an invention

programming: developing a sequence of instructions that causes a computer to perform a task or a set of operations

telegraph: a system by which messages may be sent electronically. It was widely used from the 1840s to the 1950s. The telegraph puts messages in code and sends them along wires.

template: a pattern used as a guide

transmitter: the device that changes radio waves in order to send them out as radio or TV signals

Source Notes

13 Susmitha Reddy Jeendru, "Ajay Bhatt—USB Creator," stuMagz, accessed August 8, 2021, https://www.stumagz.com/ajay-bhatt-usb-creator/.

17 N. Lee, "Annie Easley Helped Make Modern Spaceflight Possible," Engadget, February 13, 2015, https://www.engadget.com/2015-02-13-annie-easley.html.

19 Aamna Mohdin, "Gladys West: The Hidden Figure Who Helped Invent GPS," *Guardian*, November 19, 2020, https://www.theguardian.com/society/2020/nov/19/gladys-west-the-hidden-figure-who-helped-invent-gps.

25 "Internet Hall of Fame 2017—Alan Emtage," YouTube video, 5:16, posted by the Internet Society, October 19, 2017, https://www.youtube.com/watch?v=giwBCXASWVg.

Learn More

Amson-Bradshaw, Georgia. *Pioneers of Science and Technology*. Hauppauge, NY: Barron's, 2018.

EngineerGirl
https://www.engineergirl.org

Kirkfield, Vivian. *From Here to There: Inventions That Changed the Way the World Moves*. Boston: Houghton Mifflin Harcourt, 2021.

Mann, Dionna L. *Hidden Heroes in Space Exploration*. Minneapolis: Lerner Publications, 2023.

National Geographic Kids—Making History: African American Pioneers of Science
https://kids.nationalgeographic.com/science/article/black-inventors-and-pioneers-of-science

Newland, Sonya. *Working with Computers and Robotics*. Tulsa: Kane Miller, 2022.

Scratch: Starter Coding Projects
https://scratch.mit.edu/starter-projects

Trusty, Kathy. *Black Inventors: 15 Inventions That Changed the World*. Emeryville, CA: Rockridge, 2021.

Index

Photo Acknowledgments

Image credits: monkeybusinessimages/Getty Images, p. 4; Science Museum Group Collection © The Board of Trustees of the Science Museum (top), p. 5; Courtesy of the Smithsonian Institution (bottom), p. 5; John Schanlaub/Wikimedia Commons (2.0), p. 6; VTR/Alamy Stock Photo, p. 7; Alpha Historica/Alamy Stock Photo, p. 8; Ken Hawkins/Alamy Stock Photo, p. 9; JORGE TERRE OLIVA/Wikimedia Commons (SA 4.0), p. 10; ullstein bild Dtl/Getty Images, p. 11; Anthony Pidgeon/Getty Images, p. 12; Ambuj Saxena/Wikimedia Commons (2.0), p. 13; U.S. Army Photo/ARL Technical Library, pp. 14, 15; NASA/GRC, p. 16; NASA, p. 17; U.S. Army Photo by Adrian Cadiz, p. 18; U.S. Navy, p. 19; Tuan Tran/Getty Images, p. 20; © Charles W. Rogers (courtesy of Lynn Conway), p. 21; Radia Perlman/Wikimedia Commons, p. 22; GhosT/Wikimedia Commons, p. 23; Archive PL/Alamy Stock Photo, p. 24; pearleye/Getty Images, p. 25; Courtesy of Naipec, p. 26.

Cover images: Alpha Historica/Alamy Stock Photo; NASA; © Lynn Gilbert.